Collector's Guide
Three Phases of
Titania
Rutile, Anatase, and Brookite

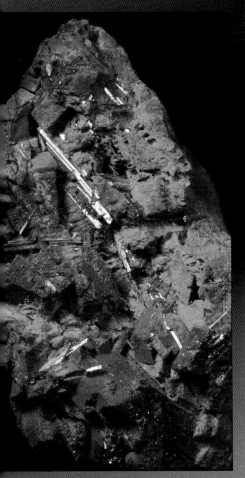

Schiffer Earth Science Monographs Volume 6

Schiffer
Publishing Ltd

4880 Lower Valley Road, Atglen, Pennsylvania 19310

Robert J. Lauf

Other Schiffer Books by Robert J. Lauf

Collector's Guide to the Axinite Group.
    ISBN: 9780764332166. $19.99.
Collector's Guide to the Epidote Group.
    ISBN: 9780764330483. $19.99.
Collector's Guide to the Mica Group.
    ISBN: 9780764330476. $19.99.
Collector's Guide to the Vesuvianite Group.
    ISBN: 9780764332159. $19.99.
Introduction to Radioactive Minerals.
    ISBN: 9780764329128. $29.95

Other Schiffer Books on Related Subjects

Collector's Guide to Fluorite. Arvid Eric Pasto.
    ISBN: 9780764331930. $19.99.
Collecting Fluorescent Minerals. Stuart Schneider.
    ISBN: 0764320912. $29.95.
Gems & Minerals. Dr. Andreas Landmann.
    ISBN: 9780764330667. $29.99.
The World of Fluorescent Minerals. Stuart Schneider.
    ISBN: 0764325442. $29.95.

Copyright © 2009 by Robert J. Lauf
Library of Congress Control Number: 2009924031

Designed by Mark David Bowyer
Type set in Arno Pro / Humanist 521 BT

ISBN: 978-0-7643-3268-5
Printed in China

Contents

Preface

This monograph is devoted to the three common titanium dioxide minerals, namely, rutile, anatase, and brookite. For completeness, mention is also made of the two very rare high-pressure phases that can be found in heavily shocked rocks associated with meteor impacts. The titanium dioxide system is perhaps unique in that all three of the common phases can be found in large, well-formed crystals at many localities around the world, making it of great interest to collectors. Rutile is of some interest to the lapidary artist as well, mainly for its role in rutilated quartz and star sapphire. These minerals are also important industrial commodities that can literally be found in some form in virtually every household. In light of the spectacular finds of anatase and brookite in Pakistan in the last five years, the author considers it timely to take a detailed look at these three minerals, not only with respect to their "classic localities," but also their fascinating diversity of forms and the examples they provide of such phenomena as oriented growth, twinning, and pseudomorphism.

The present monograph is organized as follows: After a brief introduction, the general treatment begins with an explanation of the structures of rutile, anatase, and brookite and their phase relations. A section on their formation and geochemistry explains the kinds of environments where these minerals are formed. Then, a detailed entry for each mineral provides information on important localities and full-color photos so that collectors can see what good specimens look like and how the minerals illustrate interesting mineralogical phenomena. In contrast to many encyclopedias and field guides that (because of practical limitations) attempt to present one photo as representative of a particular mineral, the specimens herein were selected to present as broad a range as possible. It is the author's belief that the collector will benefit from seeing many different examples of each particular mineral, even if some specimens are superficially similar, because each specimen has its own story to tell. As in the author's earlier volumes, the photographs were not selected to showcase extremely expensive museum pieces or purported "best in the world" specimens, but instead, to illustrate good specimens that an interested collector could actually hope to obtain and study.

Acknowledgments

The following colleagues kindly provided technical information, literature, and helpful discussions: Ahmed El Goresy, *Universität Bayreuth, Germany*; Deborah Cole, *Oak Ridge National Laboratory*; Teresa Fortney, *Oak Ridge Public Library*; Pavel Kartashov, *Institute of Geology of Ore Deposits, Russian Academy of Science*; Travis Paris; Arvid Pasto; Lisa Swain, *Huntsman Pigments*. Important specimens and background information were supplied by Laurie Adams, *The Adams Collection*; Dudley Blauwet, *Mountain Minerals*; Dave Bunk; Sharon Cisneros, *Mineralogical Research Co.*; Richard Dale, *Dale Minerals*; Jordi Fabre, *www.fabreminerals.com*; Shields Flynn, *Trafford-Flynn Minerals*; Pete Heckscher, *The Crystal Circle*; Leonard Himes, *Minerals America*; Mohammed Javed, *Javed's International Gem Imports*; Brian Kosnar, *Mineral Classics*; Rob Lavinsky, *The Arkenstone*; Gary Maddox, *Apalachee Minerals*; Tony Nikischer, *Excalibur Mineral Co.*; Neal Pfaff, *M. Phantom Minerals*; C. Carter Rich; Jeff Schlottman, *Crystal Perfection*; Jaye Smith, *The Rocksmiths*; Chris Wright, *Wright's Rock Shop*.

Introduction

The titanium oxide minerals rutile, anatase, and brookite are well known to mineral collectors and lapidary hobbyists. Many would be surprised to learn that the mining of titanium oxide, primarily for pigments, is a $3 billion industry. Titanium oxide pigments can be found in everything from plastics to paints, and even in cosmetics, flour, and toothpaste. Both rutile and anatase exhibit the high refractive index and light scattering properties that are needed to make a good white pigment. They are extremely stable against degradation by sunlight or environmental factors. For some applications, particularly in spun textile fibers, anatase (Mohs hardness 5.5) is desirable because it is less abrasive than rutile (Mohs hardness 6.5). Anatase also has a higher ultraviolet reflectance, which maximizes the effect of fluorescent whitening agents that eliminate yellowish tones and yield a pleasing bluish-white color in daylight.

All three species have been known for nearly two centuries and there is some disagreement in the literature concerning their type localities, particularly because there appear to be no "type specimens" of rutile or brookite that were formally deposited in museums. For example, Mandarino and Back (2004) give the type locale for rutile as "Revuca in Slovakia or Cajuelo (Ercajuelo), Burgos, Spain" and flatly state that type locales for anatase and brookite are unknown. However, the original description of brookite (Levy 1825) is clearly based on samples from two locales, both of which are known today. Anatase was described by early workers under a variety of names such as oisanite, dauphinite, and octahedrite. The name oisanite was applied to material from St. Christophe-en-Oisans, Isere, France, by Delametherie (1797); several years later Haüy applied the name anatase to similar material from the same general area, and type specimens of the anatase of Haüy are held in the collection of Museum of Natural History in Paris. A strong case can therefore be made that the type localities for all three species can be identified with some certainty.

Figure 1. Titanium dioxide powder and some familiar products containing titania pigments, including PVC pipe, a digital timer with injection-molded housing, and typing correction fluid. (*Pigment sample courtesy of Huntsman Pigments*).

As a gem material, natural rutile is occasionally faceted, but the stones tend to be so dark that all but the very smallest are virtually opaque. Colorless synthetic rutile ("titania") was first developed in the late 1940s and was occasionally promoted as a diamond simulant, but its optical properties (especially its high dispersion) make it easy to distinguish from real diamond and it has never caught on as a gemstone. Clean crystals of anatase can also be faceted but they, like rutile, tend to be quite dark. Lapidaries are familiar with *rutilated quartz*, working it into attractive faceted stones, cabochons, tumbled freeforms, and spheres. (A material that looks very similar to rutilated quartz is the so-called *rutilated topaz*, which was first marketed from Brazil and has more recently been found in Sri Lanka. However, it has been shown (Koivula 1987) that the golden-brown hair-like inclusions in "rutilated topaz" are not rutile crystals at all, but rather hollow tubular channels in the topaz that have become filled or coated on the inside by colloidal iron oxides.) As a gem material, rutile is most important for the supporting role it plays in creating asterism in star sapphire and star ruby.

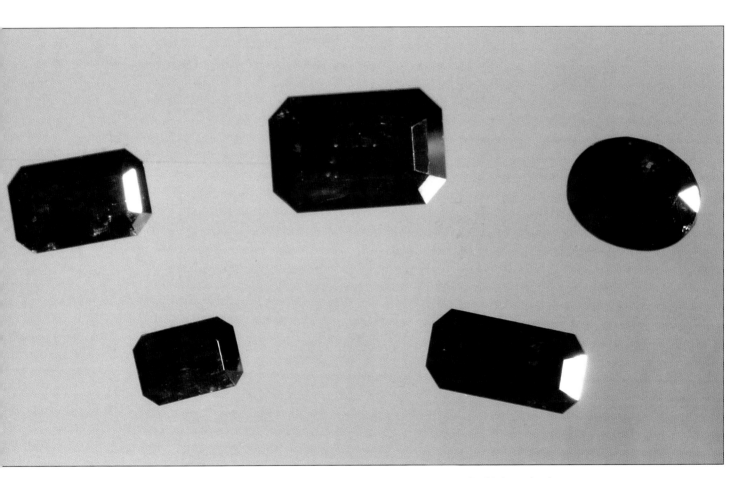

Figure 2. Faceted rutile from Zegi Mountain, Pakistan. The stones are very dark, but with strong backlighting the deep red color can be seen. *RJL3413*

Figure 3. Faceted anatase from Kharan, Baluchistan, Pakistan. As with the rutile in the previous photo, the stones are quite dark. *RJL3415*

Figure 4. A fine example of rutilated quartz crystals from Minas Gerais, Brazil. Specimen is about 11 cm tall. *RJL2477*

Figure 5. Freeform cabochons of rutilated quartz from Brazil, totaling about 100 carats. One can see that the material presents a great variety of textures for the lapidary artist to use to achieve a particular look. The color of the rutile (reddish-brown to bright gold), the size and loading of the rutile fibers (dense mats of fine needles to a few sparse but thicker crystals), and the presence of other inclusions all contribute to making each stone unique.

Figure 6. Two cabochons showing asterism caused by oriented needles of rutile in corundum; left: a 100 carat star ruby from India, right: a 4 carat black star sapphire from Sri Lanka.

Structure and Phase Relations

Crystal Structure and Morphology

In the crystal structure of rutile (Vegard 1916; Green-wood 1924; Vegard 1926) each titanium ion is sur-rounded by six oxygen ions, which form the corners of a slightly distorted regular octahedron. Each oxygen ion is surrounded by three Ti ions that lie in a plane and form a roughly equilateral triangle. The tetragonal unit cell contains two TiO_2 formula units. Rutile has a lower specific volume and therefore higher density than either anatase or brookite (Deer, Howie, and Zussman 1962).

The crystal structure of anatase was also first studied by Vegard (1916), who showed that the structure, like that of rutile, contains Ti ions surrounded by six oxygens, with every oxygen ion lying between three Ti ions. The difference is in the relative arrangements of the oxygen octahedra. In rutile, two opposite edges of each octahedron are shared with other octahedra; whereas in anatase the shared edges at the top and bottom of an octahedron lie at right angles to one another (Deer, Howie, and Zussman 1962). More recent refinements of the crystal structure have included measurements at different temperatures (Horn, Schwerdtfeger, and Meagher 1972). Anatase, like rutile, has tetragonal symmetry.

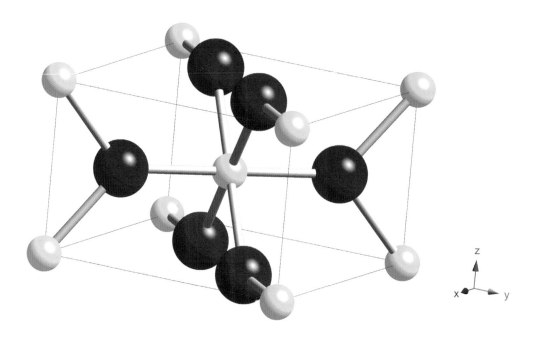

Figure 7. Ball and stick model of the rutile crystal structure. In this representation, each red ball represents an oxygen ion and each blue ball represents a titanium ion. One can see that the central Ti lies at the center of an octahedron defined by the oxygen ions at its corners. The blue lines represent the unit cell.

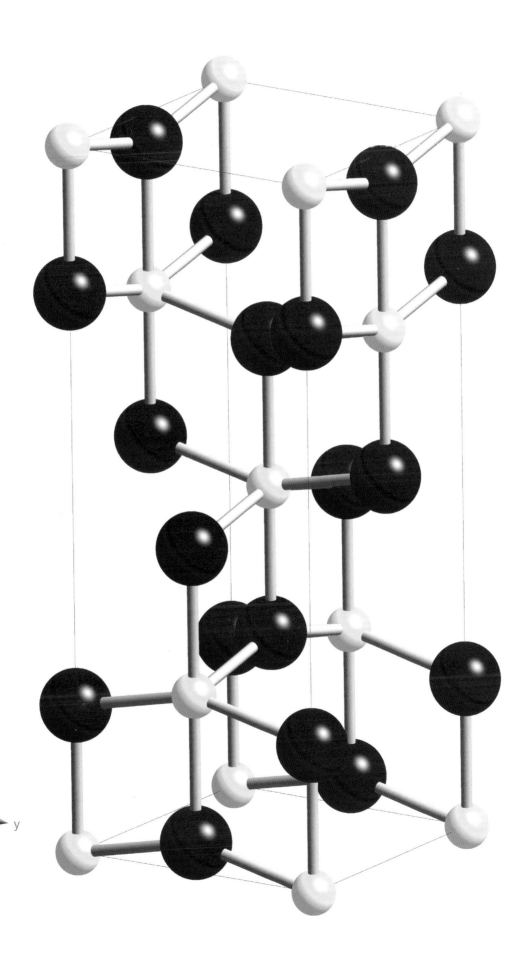

Figure 8. Ball and stick model of the anatase crystal structure.

z

x   y

The structure of brookite was first worked out by Pauling and Sturdivant (1928), who demonstrated that the structure, like those of anatase and rutile, contains octahedrally coordinated Ti, with each oxygen surrounded by three Ti ions. The main difference is that in brookite the oxygen octahedra lie in a zigzag pattern instead of in straight rows (Deer, Howie, and Zussman 1962). The resulting structure has orthorhombic symmetry.

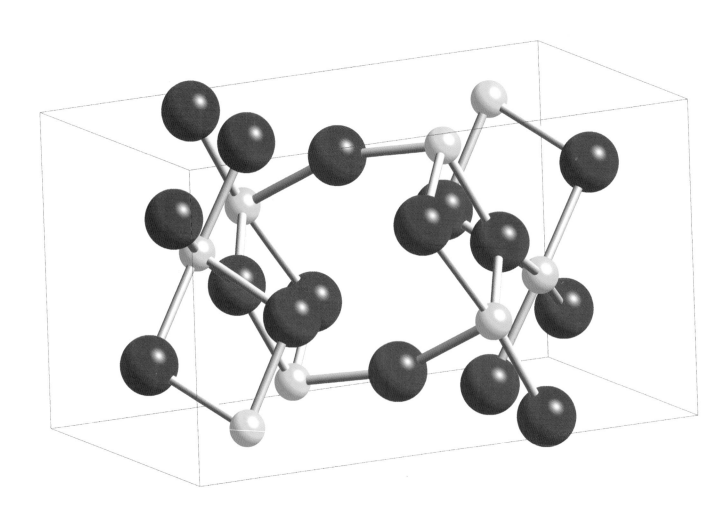

Figure 9. Ball and stick model of the brookite crystal structure.

Opposite Page:
Figure 10. Comparison of the structures of rutile, anatase, and brookite, adapted from Pauling and Sturdivant (1928). Filled circles represent Ti^{4+} ions, open circles represent O^{2-} ions. In rutile, top, the edges that are shared with adjacent oxygen octahedra (arrows) lie parallel to one another. In anatase, lower left, the shared edges at top and bottom of the octahedra lie perpendicular to one another (arrows). In brookite, lower right, the octahedra lie in a zigzag pattern.

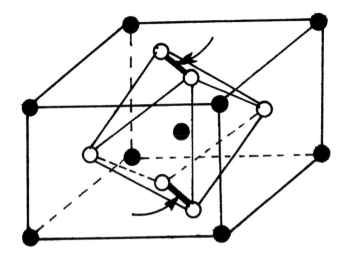

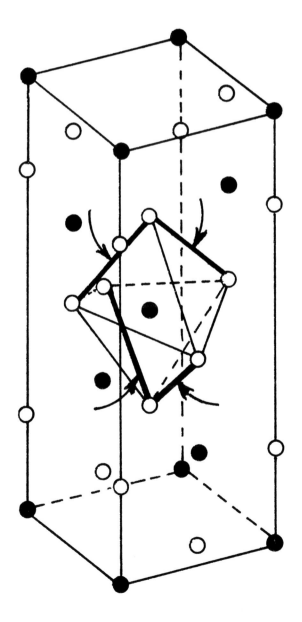

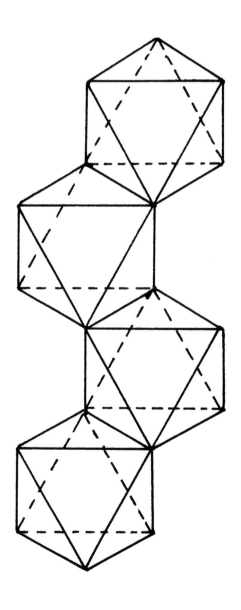

The minerals present a wide diversity of morphological forms. Goldschmidt presented 128 drawings of natural anatase crystals, 121 of brookite, and 149 of rutile.

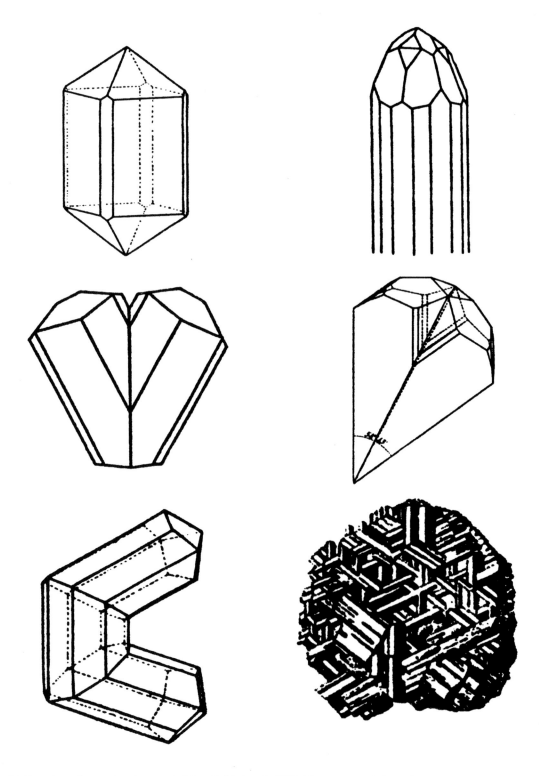

Figure 11. Drawings of natural rutile crystals (after Goldschmidt 1922). Top row: two habits of prismatic crystals; middle row: two types of v-twinning; bottom row: a cyclic twin and reticulated twinning ("sagenite").

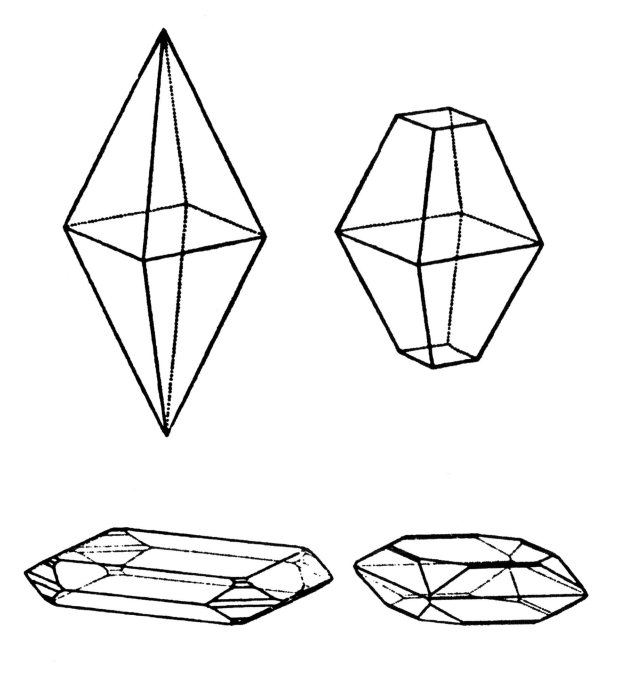

Figure 12. Drawings of anatase crystals (after Goldschmidt 1913). Upper: octahedron and truncated octahedron; lower: tabular habits.

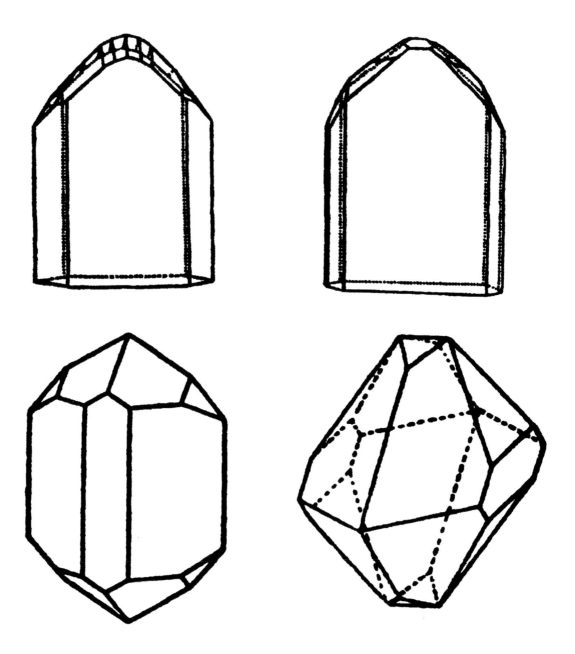

Figure 13. Drawings of brookite crystals. Upper: tabular forms from Wales (left) and France (right) illustrated by Levy (1825); lower: equant forms from Magnet Cove, Arkansas, illustrated by Goldschmidt (1913).

Phase Relationships in the TiO_2 System

Compared to other systems that exhibit polymorphism such as Al_2SiO_5 (andalusite-sillimanite-kyanite) and SiO_2 (quartz-cristobalite-tridymite), the titanium dioxide system is unique in that the three TiO_2 minerals (rutile-anatase-brookite) can all form excellent, large crystals that are sought by collectors, in relatively good abundance worldwide. At the same time, there is an interesting parallel to the silica system: both SiO_2 and TiO_2 also have very rare high-pressure phases, typically found as microscopic grains in highly shocked rocks such as those formed in the neighborhood of impact craters. The high pressure silica minerals, coesite and stishovite, have been known for over forty years, whereas high pressure titania minerals, known synthetically since the late 1960s, have only been found in nature within the last decade and are currently in the process of formal description and naming.

Another unique aspect of the TiO_2 system is that under virtually all conditions, anatase and brookite are metastable with respect to rutile. Upon heating, both minerals will transform to rutile rather than melt, and rutile will never transform back to either of these phases. According to one calculation, anatase would only be more stable than rutile under hypothetically large *negative pressures*. Attempts to study the equilibrium relationships at elevated temperatures and pressures, such as those reported by Dachille, Simons, and Roy (1968) are confounded by the sluggishness of the reactions, so it is very difficult to know if equilibrium has been reached in the time scales available in laboratory tests (Jamieson and Olinger 1969). To improve the reactivity and speed the reaction kinetics, nanoscale powders or gels are sometimes used in these experiments. Many chemical processes have been used to synthesize colloidal TiO_2 gels, such as the reaction of titanium ethoxide with ethanol or of titanium chloride and ammonium hydroxide. The precipitated gels will give X-ray diffraction patterns indicating that they are essentially amorphous. As the gels are heated, anatase or brookite may be formed; upon further heating, the material transforms to rutile after a certain particle size is reached. The rutile, once formed, grows faster than the anatase. These transformations are influenced by impurities, synthesis conditions, particle size, and the atmosphere in which the sample is heated. Because the three phases have different surface energies, it can be shown that when the particles are extremely small, brookite and/or anatase may actually be stable relative to rutile. When the particles grow large enough to overcome this surface energy difference, rutile then becomes stable and cannot be transformed back to either of the other two phases (Ranade et al. 2002). In geological processes, kinetic factors allow anatase and brookite to form at lower temperatures, even though rutile is still the more thermodynamically favored polymorph. At the same time, the energy differences between the three minerals are fairly small, a factor that explains why one can find two or sometimes all three of the minerals in close proximity. Because rutile is the most stable phase, one can find rutile pseudomorphs after anatase and, more rarely, after brookite. Rutile, on the other hand, will never transform to either anatase or brookite.

High Pressure TiO_2 Polymorphs

The high-pressure polymorphs of rutile (usually called TiO_2 II) have been synthesized by various workers and attempts have been made to clarify their stability relations at high temperatures (Withers, Essene, and Zhang 2003). Tiny amounts of these materials have been found in nature in heavily shocked garnet-cordierite-sillimanite gneiss from the Ries Crater in Germany (El Goresy et al. 2001a, 2001b).

The first mineral, **IMA 2007-058**, is monoclinic and isostructural with the baddeleyite polymorph of ZrO_2. It is the first natural occurrence of an ultradense titanium dioxide in which the Ti cations are in seven-coordinated oxygen polyhedra. It occurs as small polycrystalline aggregates of tiny ($<1\mu m$) crystallites in the parent rutile grains. Using reflected-light microscopy, the mineral may be recognized by intense blue internal reflections, in contrast to the white-to-yellow internal reflections of the surrounding rutile. The calculated density, 4.75 g/cm^3, makes it about 11 percent denser than rutile. The type material contains small amounts (0.1 to 0.2 weight percent) of iron and niobium, which might help to stabilize the structure at ordinary pressures, but residual stresses in the material might also play a role. It has been estimated that peak shock pressures ranged from 16 to 20 GPa and that post-shock temperatures were much lower than 500°C (El Goresy et al. 2001a).

The second mineral is orthorhombic with α-PbO_2 structure. The occurrence at Ries Crater marks the first observation of natural material corresponding to a high-pressure TiO_2 polymorph previously synthesized in the laboratory at pressures above 6 GPa. The mineral forms at grain boundaries between rutile and shock-compacted biotite. It is recognized by pink internal reflections in the microscope. The calculated density, 4.34 g/cm^3, makes it about 2 percent denser than rutile (El Goresy et al. 2001b). This phase has been proposed as a new mineral; currently, work is ongoing to clarify the difference between this phase and srilankite, $(Ti,Zr)O_2$, a low-pressure mineral that has the same structure but contains essential zirconium (Ahmed El Goresy, *pers. comm.* 2008).

Because of their small size and extreme rarity, these two mineral phases are of limited interest to collectors. However, they are very important to the study of mineral stability at pressures that exist in the Earth's upper mantle. Furthermore, it is interesting to note that rutile has the same structure as stishovite, a high-pressure SiO_2 phase. Therefore, an understanding of the high-pressure TiO_2 minerals can provide insights into the behavior of silica at progressively higher pressures above the stability range of stishovite, which are far too high to study directly in the laboratory (El Goresy et al. 2001a).

Formation and Geochemistry

Titania in Igneous Rocks

In magmas, titanium is present as the Ti^{4+} ion, which can substitute for Al^{3+} in six-coordination in some silicate minerals, so it can be found in pyroxene, hornblende, and biotite; however, much of the Ti tends to crystallize as tiny grains of rutile or in mixed oxides such as ilmenite ($FeTiO_3$). Large rutile crystals are occasionally formed in granite pegmatites and in apatite and quartz veins. In the quartz-feldspar veins at Mount Kapudzhukh, Azerbaijan, rutile is associated with wolframite and scheelite. Red, prismatic rutile is common in pegmatite at Hiddenite, North Carolina, associated with muscovite, beryl, and occasionally other minerals (Brown and Wilson 2001).

Rutile can make up as much as 1 to 5 percent of certain uncommon rocks such as nelsonite and kragerøite (Deer, Howie, and Zussmann 1962). Nelsonites are a type of nonsilicate ultramafic rock typically composed of about two-thirds by volume of Fe-Ti oxides and one-third apatite (Middlemost 1985). Krageøite is a rutile albitite rock found at Lindviksollen, near Kragerø, Norway; it is a granular metagabbro consisting mainly of plagioclase feldspar (albite), hornblende, and several percent rutile.

Figure 14. Complex black rutile crystals associated with terminated yellowish quartz on slightly porous matrix, typical of the vein deposits at Mount Kapudzhukh, Azerbaijan. *RJL2348*

Figure 15. Prismatic rutile crystals on muscovite from North America Emerald Mines, Hiddenite, North Carolina, typical of a pegmatite environment. *RJL3316*

Figure 16. A less common association: prismatic rutile crystals penetrating an emerald crystal, from North America Emerald Mines, Hiddenite, North Carolina. The emerald is about 1 cm across. *RJL3422*

At Zegi Mountain, Pakistan, crystals are found in a highly corroded alkaline pegmatite, which is so thoroughly decomposed that the more resistant phases, such as rutile, can in some cases be dug out with bare hands. At the Agua Colgada quarry, Atacama Region, Chile, rutile is found in pegmatite veinlets that contain a type of diorite rock called "leucotonalite."

The niobium- and tantalum-rich varieties of rutile, *ilmenorutile* and *strüverite*, are typically found in pegmatites.

Some notable localities include: the McGuire pegmatite, Park County, Colorado; the Etta and Peerless mines, near Keystone, South Dakota; at Udraz, Czech Republic; near Craveggia, Val Vigezzo, Italy; at Miass, Southern Ural Mountains, Russia; in Madagascar at Ambatofinandrahana, Ampangabe, and Antsakoa. Large strüverite crystals have recently been reported from the Borborema pegmatitic province in Rio Grande do Norte, Brazil (Beurlen et al 2004).

Figure 17. Typical striated prismatic rutile crystal about 25 mm tall, with minor albite, recovered from corroded pegmatite at Zegi Mountain, Pakistan. *RJL3359*

Figure 18. Dark red rutile needles in bundles and sagenitic masses, with white feldspar from a pegmatite (leucotonalite) veinlet at the Agua Colgada quarry, Freirina, Atacama Region, Chile. Specimen is about 6 cm tall. *RJL3431*

Anatase is occasionally found in granitic pegmatites, and has been reported from cavities in andesite at Komna, Moravia, Czech Republic. It is a minor accessory mineral in the Dartmoor granite, Devonshire, England. Anatase is "very widespread" at Mont Saint-Hilaire, Quebec, Canada, where it has been found in pegmatites, altered pegmatites, syenites, and other rocks; rutile and brookite are also found there, in lesser amounts (Mandarino and Anderson 1989).

Figure 19. A minute black crystal of brookite, about 0.3 mm across, collected by Dave Richerson at Mont Saint-Hilaire, Quebec, Canada, in the 1970s.

Brookite (along with lesser amounts of anatase and rutile) is found in alkaline syenite at Magnet Cove, Arkansas. It occurs with quartz crystals at the Ellenville Lead mine, Ulster County, New York. Minute tabular brookite crystals with black bipyramidal anatase can be found in corroded granite at Topsham, Maine.

Figure 20. A thumbnail-sized specimen of submetallic blue-black crystals of brookite on milky quartz from Magnet Cove, Arkansas. *RJL664*

Figure 21. A sample of corroded graphic granite from Topsham, Maine, containing dark microcrystals of anatase and brookite. Sample is about 3 cm wide. *RJL211*

Figure 22. Detail of specimen in the previous figure showing tabular brookite crystal about 0.5 mm long, in feldspar. *RJL211*

Titania in Metamorphic Rocks

Rutile is a common accessory mineral in many metamorphic rocks, particularly eclogites and amphibolites. Small acicular rutile crystals are sometimes found in reconstituted sediments such as clays and shales and also in zones where these sediments have undergone contact metamorphism (Deer, Howie, and Zussmann 1962). The conversion of ilmenite to mixtures of rutile and hematite, magnetite, or siderite is attributed to metamorphic processes.

Complex metamorphic processes are responsible for important occurrences of fine rutile crystals in pyrophyllite at Graves Mountain, Georgia, and at the Champion andalusite mine, Mono County, California (Cook 1985, Cook 2003a; Wise 1977).

The Onganja mining district, Namibia, perhaps more noteworthy to collectors for fine copper minerals, has also yielded large (> 10 cm) prismatic rutile crystals. The crystals are typically well terminated, and their prism faces are usually striated parallel to the *c*-axis. The geology is interpreted as fracture-filled veins that seem to be tensional features cutting through gneiss and mica schist (Cairncross and Moir 1996).

Figure 24. Dark red, striated complex rutile crystal about 3 cm tall, from the Onganja mine, Namibia. *RJL3311*

Figure 23. A thumbnail-sized specimen with several small, deep red rutile crystals in beige pyrophyllite, from Graves Mountain, Georgia. The deposit is believed to be the result of both hydrothermal alteration and regional metamorphism. *RJL716*

Many of the finest anatase and brookite specimens are found in Alpine-cleft type deposits. These formations develop in small fissures or tension cracks in which a slow-moving hydrothermal fluid infiltrates and recrystallizes the adjacent wall rock. The fluids themselves are thought to be derived from nearby rocks through regional metamorphic processes. Cleft deposits tend to be small and because the space is confined, removal of specimens is a very painstaking process. Alpine-cleft occurrences include many classic locales. The original description of brookite (Levy 1825) was based on specimens from Tremadoc, Wales, and Dauphin, France, both of which are localities with cleft-type deposits. Other cleft deposits include: Hardangarvidda, Hordaland, Norway; Prägratten, Austria; numerous locales in Switzerland; at Passo di Vizze, Trentino-Alto Adige, and at Ghiacciaio Miage, Val d'Aosta in Italy; and the quartz veins at Dodo, Subpolar Urals, Russia. The superb anatase and brookite specimens being found in Pakistan are coming from a number of Alpine-cleft locales in Baluchistan. The fine specimens of rutilated quartz and oriented rutile on hematite from near Ibitiara, Bahia, Brazil, are also examples of cleft-type deposits.

Figure 25. A tiny bi-pyramidal anatase crystal on matrix, from the type locale at Bourg d'Oisans, France. *RJL3364*

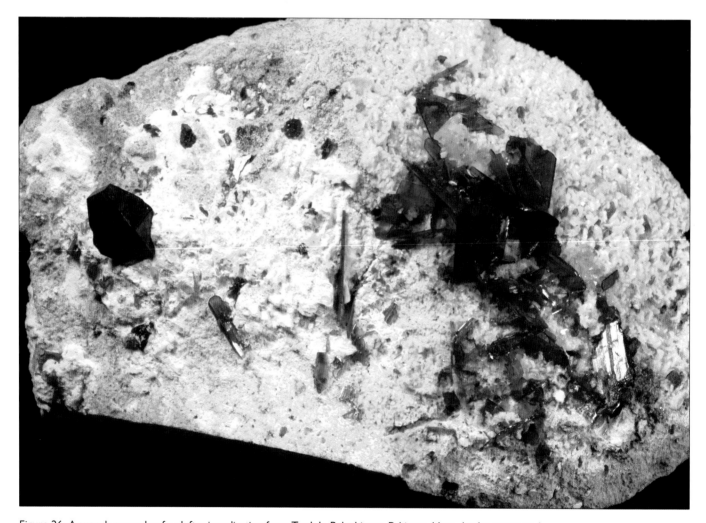

Figure 26. A superb example of a cleft mineralization from Trodok, Baluchistan, Pakistan. Here, both anatase and brookite occur within a few centimeters of each other. One can speculate that the temperature (or possibly other variables) changed as the mineralization developed, allowing one of the phases to crystallize first and then shifting to a regime in which the second phase crystallized. Because the two minerals are not very different energetically, and kinetics are sluggish, the earlier-formed mineral did not re-dissolve or transform to the other phase. *RJL2973*

Figure 27. A 2-cm brookite crystal attached to the side of a rough quartz crystal associated with minor green chlorite, from the quartz vein deposit at Dodo, Russia. *RJL2687*

Figure 28. Dark blue anatase crystals to about 3 mm in chlorite schist, from Dodo, Russia. *RJL1309*

Figure 29. Deep blue anatase crystal about 1 cm long on a large colorless quartz crystal, from Diamantina, Minas Gerais, Brazil. With backlighting, the anatase is actually transparent. *RJL3408*

Figure 30. A specimen that might be regarded as more interesting than beautiful: dark blue anatase crystals sprinkled on a large smoky quartz crystal, which is heavily included with micaceous green chlorite, illustrating a typical cleft-type association. The specimen is about 10 cm tall and was found at Lapcha, Northern Urals, Russia. *RJL1900*

Figure 31. Minute anatase crystals on colorless quartz, from the Shingletrap Mountain mine, Montgomery County, North Carolina. Note the color zoning, particularly in the uppermost anatase crystal: darker at the midsection and translucent brown toward the terminations. *RJL3429*

Figure 32. A fine example of rutilated quartz that also includes tabular black hematite, from Novo Horizonte, Bahia, Brazil. Specimen is about 3 cm tall. *RJL1540*

Detrital Deposits

Titanium oxides are extremely stable compounds and are fairly hard, so it is not surprising that as their host rocks undergo weathering and erosion these minerals become liberated and may form part of the soil in the vicinity of a decomposing granite or pegmatite. In one early study, rutile, anatase, and brookite recovered from soils and stream sands in Dartmoor, England, were shown to be derived from the Dartmoor granite (Brammall and Harwood 1923). At a locality near Georgetown, California, brookite and anatase are found in lateritic soils that were derived from the weathering of shales containing intrusive quartz veins (Hadley 2000). Extremely small crystals of anatase (0.05 to 0.2 µm) were found to occur in sedimentary kaolins from Georgia in an early study using electron microscopy (Nagelschmidt, Donnelly, and Morcom 1949). Small, well-formed rutile crystals are abundant in the soil formed from highly weathered gneiss and schist at the emerald mines at Hiddenite, North Carolina; these can be most efficiently recovered by washing to remove the clay and "panning" to concentrate the denser rutile.

If running water is present, the minerals may be carried great distances and concentrated in alluvial deposits. In some cases, the degree of concentration is so high that it is economical to mine the Ti-rich sands by dredging on a large scale. Rutile is found in heavy mineral sands at many localities, including: Duval and Clay Counties, Florida; Georgia; the McNairy sand, which outcrops in six counties in western Tennessee; North Stradbroke Island, Australia; Sri Lanka; Cameroon; and Sierra Leone. Anatase is found as a detrital mineral in sediments, where it is sometimes presumed to have developed in place during or after deposition (a so-called authigenic origin). Detrital brookite is a common component of grits and sandstones and is also found in placer deposits being worked for diamonds or gold (Deer, Howie, and Zussman 1962). Small tabular brookites are found at Hale Creek, California, where they have weathered out of a massive chert zone that cuts across the creek (Dunning and Cooper 2000).

Figure 33. Samples of rutile sands recovered from detrital deposits. Left: rutile sand from Melbourne, Florida. Right: rutile sand concentrates from Queensland, Australia.

Figure 34. A minute, tabular brookite crystal, about 2 X 4 mm, from Georgetown, California, derived from shale by weathering. *RJL3368*

Figure 35. Detrital material recovered at the Rist mine (North America Emerald Mines), Hiddenite, North Carolina, in the course of emerald mining. A pail full of this material is estimated to contain over *nine hundred* individual crystals.

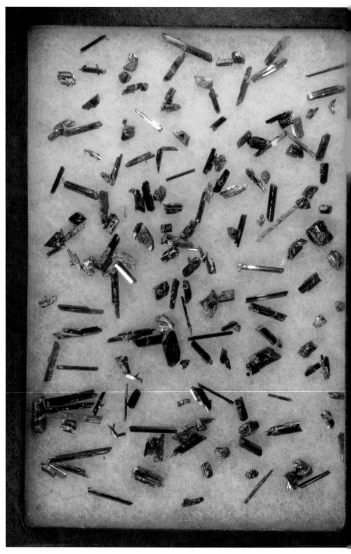

Figure 36. Riker mount showing a variety of rutile crystals recovered from the type of material shown in the previous figure by careful washing and panning. *RJL3199*

Titania in Extraterrestrial Rocks

Rutile is not normally considered to be a major constituent of meteorites. However, microscopic studies have shown it to be a fairly widespread but minor accessory mineral in a number of meteorites. Some meteorites that have been found to contain rutile include: the Allegan County, Michigan, olivine-bronzite chondrite H5; the Bondoc Peninsula, Phillipines, mesosiderite; the Estherville, Emmet County, Iowa, mesosiderite; the Farmington, Washington County, Kansas, olivine-hypersthene chondrite L5; and the Vaca Muerta, Taltal, Atacama, Chile, mesosiderite. In the Vaca Muerta meteorite the rutile is associated with chromite in the form of both individual grains and lamellae within the chromite. Lamellae of chromite are also found within large rutile grains. In contrast, the Allegan meteorite contains rutile disseminated as grains within the silicate matrix rather than as lamellae in chromite (Buseck and Keil 1966). Anatase was detected in the Say al Uhaymir 060 shergottite using micro-raman spectroscopy (Hochleitner et al. 2004). Even though these phases are very minor constituents in meteorites, interest in refractory oxides in these objects is driven in part by the desire to gain insights into the primitive state of the solar nebula, in which TiO_2 is a possible constituent (Posch et al. 2003). On the moon, niobian rutile has been found in lunar samples from both the Apollo 14 and Apollo 17 missions (Steele 1975).

Figure 37. Cut surface of the Vaca Muerta mesosiderite, Atacama, Chile, one of the meteorites in which rutile has been reported. Sample is about 5 cm wide. *RJL1218*

The Minerals

Of the three common minerals, rutile is the most abundant in nature and most familiar to collectors, but well-crystallized anatase and brookite are also known from numerous localities worldwide and deserve a place in any systematic or display collection. The three species are also of interest for specialty collections, providing excellent examples of twinning, oriented growths, inclusions, pseudomorphism, and other interesting phenomena. A single volume cannot do justice to the huge number of documented localities for rutile (>2600), anatase (>1200), and brookite (>400); in the sections that follow, preference is given to localities that are most likely to be familiar to collectors, from which specimens are reasonably available, or that illustrate important geological or mineralogical principles. The interested reader can find many references that list each particular author's selection of notable localities (e.g., Palache, Berman, and Frondel 1944; Gaines et al. 1997; Bernard and Hyršl 2004). Exhaustive locality listings are also available through various Internet resources.

Rutile

Rutile is a high-temperature mineral of wide distribution in metamorphic rocks, particularly gneiss and schist, where it can be found as a "rock forming mineral" in small individual grains within the rock itself, or as well-formed crystals in cleft fillings. It can form in crystalline limestones as a product of regional or contact metamorphism. Large crystals are found in quartzite, associated with kyanite, pyrophyllite, and other species. It is also an accessory mineral in granites and pegmatites. Because of its high chemical stability, rutile is a common detrital mineral, sometimes accumulating in economically significant quantities.

Rutile was described over two hundred years ago from a pegmatite in Spain. Although no formal "type specimen" of rutile has been preserved, the noted Spanish mineralogist

Miguel Calvo has provided some useful historical insights (Jordi Fabre, *pers. comm.* 2008): Rutile was originally described as a mineral species by Werner in 1803, based on specimens that supposedly came from "Cajuelo, Vuitrago, Burgos." For some time the mineral was even called "cajuelite" after the supposed type locality. It turns out that the real locality of the material studied by Werner was evidently Horcajuelo de la Sierra (Madrid), where it had been known as a curiosity for a long time before it was described scientifically. Consequently, Horcajuelo de la Sierra, Madrid, can be considered the type locale. The reference to "Buitrago" in the original description alludes to the historical fact that the town was part of the Dominion of Buitrago, of the Dukes of the Infantado, until its abolition in the 1830s.

Figure 38. Crude rutile crystal about 1 cm across, probably twinned, from the pegmatite at Horcajuelo de la Sierra, Madrid, Spain; this occurrence is regarded by some authorities as the type locale for rutile. *RJL3358*

The premier American locale for spectacular rutile crystals is Graves Mountain, Georgia, where lustrous crystals to several kilograms are found associated with pyrophyllite, lazulite, quartz, and other species. The deposit is interpreted as a regionally metamorphosed hydrothermal alteration zone (Cook 1985, Cook 2003a). Fine crystals to six centimeters occur in a somewhat similar mineral assemblage in metamorphosed rhyolite at the Champion andalusite mine, Mono County, California (Wise 1977). Large rutile crystals are also found at the Chubb Mountain kyanite prospect, near Lincolnton, Gaston County, North Carolina, associated with kyanite, pyrophyllite, quartz, andalusite, and lazulite.

Figure 39. Rutile crystal about 4 cm tall from Graves Mountain, Georgia. *RJL3313*

Figure 40. Equant rutile crystals, about 8 to 10 mm, on massive to botryoidal hematite, from Graves Mountain, Georgia. *RJL3424*

Figure 41. A tabular rutile crystal about 1 cm across, in pearly, beige pyrophyllite from the Champion mine, Mono County, California. *RJL376*

Figure 42. Rough intergrown rutile crystals forming a group about 3 cm across, from the Chubb Mountain kyanite prospect, Gaston County, North Carolina. *RJL3405*

At the Slyudorudnik mine, near Kyshtym in the Southern Urals, Russia, interesting large, deformed rutile crystals have been found in granular quartz. The quartz is frequently iron-stained or limonitized near the rutile crystals. The mine is a mica deposit that is now worked primarily for high-purity quartz (Evseev 1993; Pavel Kartashov, *pers. comm.* 2008). Sharp, equant crystals on colorless quartz have been mined in Azerbaijan since the late 1990s at Mount Kaputdzhukh, in the Zangezurskii Range, near the border with Armenia (Evseev 1996, Kouznetsov 2001).

Figure 43. Deep red to black rutile with a metallic luster, in iron-stained granular quartz, from the Slyudorudnik mine, near Kyshtym, Southern Urals, Russia. The crystals show significant deformation characteristic of samples from this mine. The crystals are about 3 cm tall. *RJL3261*

Figure 44. Lustrous black 15-mm rutile crystal in a quartz-lined vug from Mount Kapudzhukh, Azerbaijan. *RJL2441*

North Carolina is notable for ruby-red prismatic crystals found in a variety of geologic settings, each with its own unique associations. At Hiddenite, for example, crystals are associated with muscovite and are also found in huge numbers as small, loose crystals in mud. Near Boiling Springs, Cleveland County, reticulated crystal groups are found that have grown in an oriented pattern on rhombs of siderite or ankerite, which were later altered to goethite.

Figure 45. Gemmy, red, elongated prismatic rutile crystals about 8 mm long, on muscovite from near Shelby, Cleveland County, North Carolina. *RJL3317*

Figure 46. Red elongated rutile prisms, oriented on goethite pseudomorphic after siderite(?) from Cleveland County, North Carolina. Specimen is about 7 cm tall. *RJL2425*

Figure 47. Another example from the same locality, showing reticulated needles of rutile that seem to have a relationship to the underlying matrix. *RJL3423*

A good review of several dozen localities in the United States and worldwide was recently published by Cook (2003a).

Inclusions in Other Crystals

Most collectors and lapidaries are familiar with *rutilated quartz* from Brazil. It is interesting to note that in this material the rutile needles grew first into an open cleft or cavity and were later engulfed by a growing quartz crystal, so the relationship between the two minerals is random; in contrast, the rutile in star sapphire forms by an exsolution process along particular crystallographic planes, creating the six-rayed asterism when viewed along the *c*-axis of the corundum crystal.

Figure 48. A crude rutile crystal from Vojslavice, Czech Republic, about 2 cm wide, showing some abrasion and surface alteration, suggesting that it was removed from its original matrix by weathering. *RJL3437*

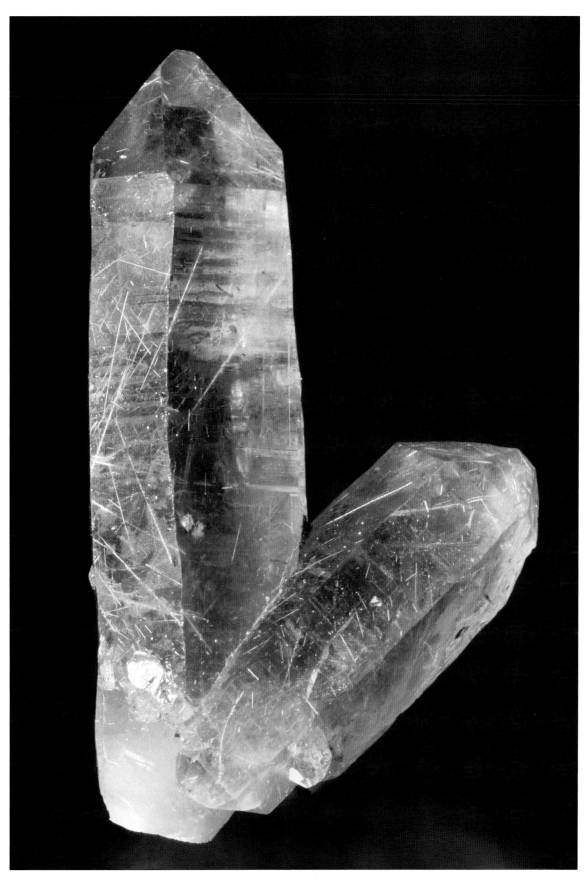

Figure 49. A rutilated quartz crystal group about 10 cm tall, from Ibitiara, Bahia, Brazil. *RJL2312*

Figure 50. A thin plate, about 3 X 6 cm, covered with colorless quartz crystals enclosing golden rutile needles, from Novo Horizonte, Bahia, Brazil. At several points on this sample, rutile crystals pass into or emerge from a particular quartz crystal, or pass directly from one crystal into another, demonstrating that the rutile did not grow inside the quartz through a precipitation process, but rather was formed first and later enclosed by the growing quartz. *RJL3047*

Figure 51. A rutilated quartz crystal showing very clearly that a quartz crystal engulfed a preexisting assemblage comprising black tabular hematite thickly covered by epitaxial growths of hair-like rutile, from Novo Horizonte, Brazil. *RJL3412*

Figure 52. An even more complicated example, in which silvery-gray fibrous rutile grew on thin tabular brookite crystals, after which the entire assemblage was overgrown by a large colorless quartz crystal. The quartz is about 5 cm tall and comes from Kharan, Baluchistan, Pakistan. *RJL2972*

Oriented Growth

Rutile has been found as oriented overgrowths on hematite, magnetite, ilmenite, brookite, and anatase. Microscopic arrays of oriented rutile fibers are observed in corundum, phlogopite, pseudobrookite, and quartz, and are the cause of asterism in star sapphire as well as some phlogopite and quartz. Spectacular examples of sharp, golden, radiating needles thickly growing on black tabular hematite are found at Ibitiara, Bahia, Brazil. An interesting find of oriented rutile crystals interspersed with fine-grained hematite was made in 2007 in the Mwinilanga region of Zambia. The assemblage is thought to be a pseudomorphic replacement of ilmenite under hydrothermal conditions, with the rutile having an epitactic relationship to the original ilmenite crystal.

Figure 53. Golden rutile needles growing in an oriented arrangement on black hematite from Ibitiara, Brazil. Field of view is about 5 cm wide. *RJL2453*

Figure 54. A fine example of black tabular hematite crystals, each with golden rutile needles growing from the edges. A few colorless quartz crystals have later grown on top of the assemblage. Although collectors normally see individual disk-like specimens, with only a single tabular hematite crystal, this specimen shows that the material usually forms three-dimensional clusters, which are later broken apart by miners or dealers into individual pieces for sale and display. Specimen is about 6 cm across. *RJL3425*

Figure 55. Deep red rutile crystals in an oriented arrangement surrounded by earthy black hematite, apparently replacing a tabular ilmenite crystal, from Mwinilunga, Zambia. *RJL3258*

Twinning

Rutile provides many interesting examples of twinning; in fact, of the 149 examples of natural rutile crystals illustrated by Goldschmidt (1922), roughly fifty drawings represent some form of twinning. The most commonly seen types include various forms of V-twins, "geniculated" or knee-like twins, cyclic twins, and reticulated groups commonly referred to as sagenite.

Figure 56. Reddish-black rutile with submetallic luster in what appears to be an oriented relationship on underlying hematite, associated with massive colorless quartz, from Diamantina, Minas Gerais, Brazil. Sample is about 5 cm wide. *RJL3362*

Figure 57. A fairly typical example of a rutile V-twin, 25 mm tall and 20 mm across at its widest point, from Diamantina, Brazil. *RJL2282*

Figure 58. An example of reticulated twinning, called sagenite: deep red to copper-colored rutile on massive colorless quartz from Alchuri, Shigar Valley, Pakistan. *RJL3036*

Figure 59. A small but very sharp reticulated twin about 15 mm tall, from Diamantina, Brazil. *RJL3314*

Figure 60. An example of geniculated twinning in a blood-red rutile crystal about 15 mm wide, from Cousineau, Quebec, Canada. *RJL3367*

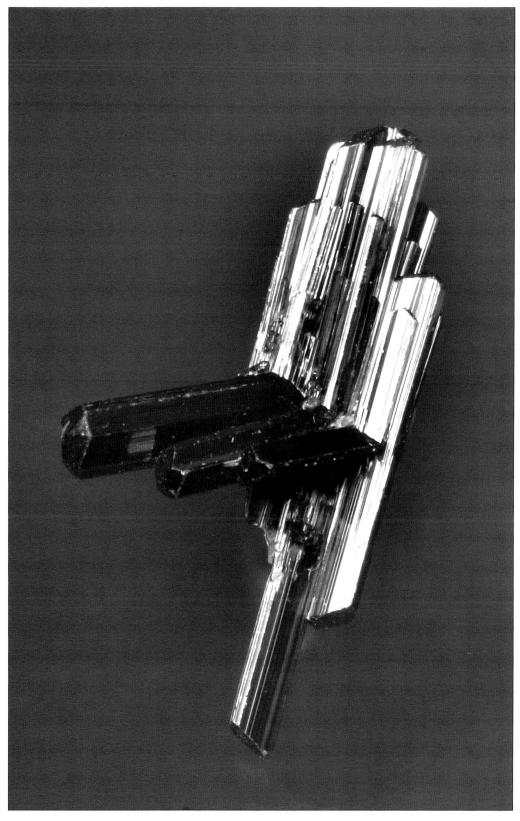

Figure 61. Another geniculated twin, about 2 cm tall, from Buritis, Diamantina, Brazil. Although the crystals have a more elongated habit than seen in the preceding photo, the angle formed at their twin plane is identical. *RJL3329*

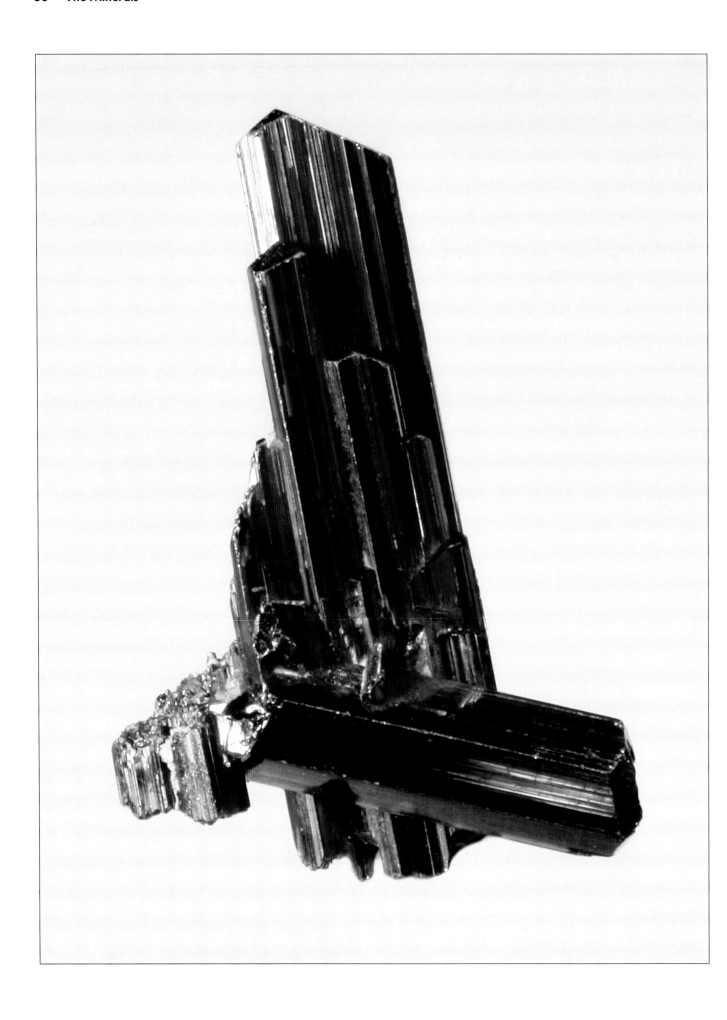

Figure 63. A large twinned rutile crystal about 3 cm wide, from an abandoned quartz quarry on Herzogberg Mountain, near Modriach, Styria, Austria. *RJL3409*

Opposite Page:
Figure 62. A thumbnail-sized group of rutile crystals from North Carolina, showing geniculated twinning similar to that in the preceding photo. *RJL3310*

Pseudomorphs and Alteration Products

Pseudomorphs of rutile after anatase are found at Col de la Madeleine, Maurienne, France, and at Diamantina and Gouveia, both in Minas Gerais, Brazil. Pseudomorphs of rutile after brookite occur at Magnet Cove, Arkansas.

Compositional Varieties

Several compositional variants are well known, particularly Fe-, Nb-, and Ta-rich rutile. Natural rutile almost always contains some Fe, typically ranging from a few to over ten weight percent Fe^{3+}, giving it a red to brownish-red color, the depth of which is related to the iron content.

The varietal name *nigrin* has been applied to black rutile, in which the Fe^{3+} content may be as high as 11 percent. The niobian and tantalian varieties, *ilmenorutile* and *strüverite*, respectively, tend to be dark brown to green and often opaque. Collectors should note that some authors treat ilmenorutile and strüverite as valid or arguably valid species; however, even the most Nb- or Ta-rich examples are still Ti-dominant, i.e., Ti > Nb + Ta. There is, therefore, no logical or structural basis for defining the species boundaries and both ilmenorutile and strüverite have been discredited by the IMA and are now regarded as varieties of rutile (Burke 2006). The reason these particular elements are taken up by the rutile crystal is that the ionic radii of Fe^{3+}, Nb^{5+}, and Ta^{5+} are very close to that of Ti^{4+}, and Fe^{2+} is slightly larger.

Figure 64. A heavily etched quartz crystal about 15 mm tall, with a golden-brown, somewhat silky 4-mm mass on the right side, from Diamantina, Brazil. The mass was an equant anatase crystal that was later transformed to needles of rutile. *RJL1192*

Figure 65. Scanning Electron Microscope (SEM) photo of the golden crystal in the previous figure.

When only Fe^{2+} or Fe^{3+} substitutes for Ti^{4+}, charge compensation involves oxygen vacancies; i.e., the composition would correspond to $(Ti^{4+}_{1-x}Fe^{2+}_x)O_{2-x}$. Ta^{5+} and Nb^{5+} are typically compensated via the coupled substitution of Fe^{2+} and sometimes Mn^{2+} so the resulting compositions show a characteristic relationship $(Fe+Mn)/(Nb+Ta) > 0.5$. In mine concentrates from the Orapa kimberlite, Botswana, small (5 to 10 millimeter) black nodules contain lamellar intergrowths of ilmenite and niobian rutile in which the rutile contains significant concentrations of both niobium (6.5 to 20.9 weight percent Nb_2O_5) and chromium (5.2 to 8.2 weight percent Cr_2O_3). The structure is interpreted to be the result of an exsolution-like process, but the identity of the precursor mineral is not known with certainty (Tollo and Haggerty 1987). Niobian rutile intergrown with titanian columbite has been reported from the Huron Claim pegmatite, southeastern Manitoba, Canada; the composition of that rutile roughly corresponded to $(Ti_{0.7}Nb_{0.15}Fe_{0.13})O_2$ (Černý et al. 1981).

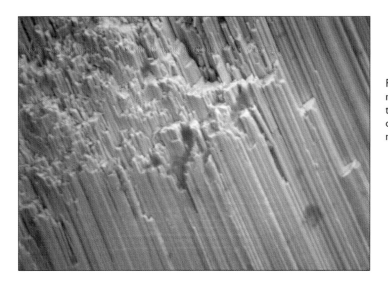

Figure 66. SEM photo at higher magnification, clearly showing that the altered crystal consists of a bundle of parallel rutile needles.

Figure 67. Elongated bi-pyramidal anatase crystals about 15 mm long, in the process of alteration to velvety rutile. The association with pale smoky quartz illustrates a typical Alpine-cleft type occurrence, from Col de la Madeleine, France. *RJL2382*

Some documented localities for ilmenorutile are Magnet Cove, Arkansas; near Lone Pine, Inyo County, California; in pegmatites at Evye, Iceland, at numerous localities in the Iveland pegmatite district, Norway, at the type locale near Miask, Ilmeny Mountains, Russia, and at Graveggia, Piedmont, Italy; at several localities in Madagascar; near Mokota, Zimbabwe; and the Kuala Kangsar district, Perak, Malaya. The material from Malaya has an interesting history: It was discovered in 1909 and originally identified as struverite (Crook and Johnstone 1912; Scrivenor 1912), and the analysis also indicated appreciable amounts of tin. Critical reexamination of this material (Flinter 1959) suggested that the original analysis was erroneous, partly because of contamination by cassiterite and ilmenite, and partly because of the difficulty of accurately measuring Nb/Ta in the early 1900s. Flinter's analysis indicated Nb:Ta ratios around 2:1, clearly placing the samples in the compositional range of ilmenorutile.

Figure 68. SEM photo of a 4-mm crystal of niobium-rich rutile var. *ilmenorutile* from Inyo County, California. *RJL1536*

Figure 69. Equant black crystals of rutile var. *ilmenorutile* to 1 cm on drusy quartz, from the Khibiny Massif, Kola Peninsula, Russia. *RJL3361*

Figure 70. An old classic: a rough single crystal of rutile var. *ilmenorutile* collected at Iveland, Norway, by A. Thortveit in November 1929. *RJL3435*

Strüverite was first described from a pegmatite near Craveggia, Val Vigezzo, Piedmont, Italy (Prior and Zambonini 1908), where it was found in small crystalline masses embedded in the quartz and feldspar. Crystals up to 8 millimeters were found and the material is black and opaque even in thin sections. The variety has also been reported from: Fefena and Ampangabe, Madagascar; and the Canoas pegmatite, Rio Grande do Norte, Brazil. At the Etta mine, Keystone, South Dakota, aggregates of small black strüverite crystals are found with muscovite in feldspar; at the nearby Peerless mine, bluish-gray crystals with a brilliant metallic luster are associated with beryl (Roberts and Rapp 1965).

Varieties of rutile containing significant amounts of other transition metals have been described from numerous localities. In many cases the rutile takes the form of small, disseminated grains rather than large "collector" specimens. Nevertheless, these materials can be important to geochemists and exploration geologists because of their association with other economically important minerals. For example, vanadium-rich rutile and other V-rich minerals are found in the epithermal gold deposit at Tuvatu, Fiji, associated with calaverite, roscoelite, vanadian muscovite, and karelianite. The presence of V-rich minerals is "consistent with the derivation of V from the alkalic intrusive rocks, which are also considered to be the source of Au and Te in the Tuvatu deposit," (Spry and Scherbarth 2006). At the Deadhorse Creek diatreme complex, northwest Ontario, Canada, minute grains of Cr-V-Nb rutile have been described in which the Nb content (up to 32% Nb_2O_5) is among the highest reported for *ilmenorutiles*. Microprobe analysis indicated significant vanadium enrichment (up to 14% V_2O_3) and the authors presented a detailed argument that in these unusual rutiles the vanadium might exist as both V^{3+} and V^{4+}. The rutile was postulated to have formed from

Figure 71. Black prismatic crystal of tantalum-rich rutile var. *strüverite*, about 15 mm tall, in massive quartz from Samiresy, Madagascar. *RJL3352*

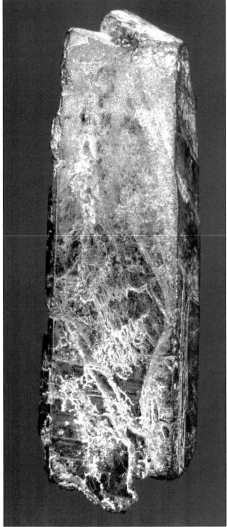

Figure 72. Simple black tabular crystal of rutile var. *struverite*, about 5 cm tall, from Tongafeno, Madagascar. *RJL3436*

hydrous alkaline solutions that had scavenged the transition metals from mafic or ultramafic source rocks (Platt and Mitchell 1996). Vanadium-rich rutiles have also been reported from: Lasamba Hill, Kenya; Balkanskoye W-Mo deposit, South Urals; Borisovskie Sopki, Plast, South Urals; and the Titova corundum pit, near Byzovo village, Middle Urals. An Fe-W rutile occurs at the Chartach scheelite deposit, Izoplit village, near Ekaterinburg, Russia.

Anatase

Anatase is characteristically found in Alpine cleft-type deposits in gneiss or schist, associated with quartz, rutile, brookite, orthoclase var. adularia, hematite, and chlorite.

Less commonly, it is an accessory mineral in some igneous rocks. Anatase is a common detrital mineral.

Many authors regard St. Christophe-en-Oisans, Bourg d'Oisans, Isère, Rhône-Alpes, France, as the type locality for anatase based on a description of the mineral in 1801 by Haüy with type specimens deposited at MHN-Paris [see the Catalogue of Type Mineral Specimens, maintained by the IMA Commission on Museums]. A brief summary of its properties was published even earlier under the name "oisanite" (Delametherie 1797). For many years the mineral was also called octahedrite in allusion to the steep bi-pyramidal crystals. Eventually, "octahedrite" was unanimously rejected in favor of anatase (International Mineralogical Association 1962).

Figure 73. Small black crystalline masses of rutile var. *struverite* in pegmatite from the Etta mine, Keystone, South Dakota. *RJL3434*

Sharp crystals are found in Alpine cleft-type deposits at a number of locales in Europe, including: Bourg d'Oisans, Isere, and near La Grave, Hautes-Alpes in France; at Tavetsch, Graubünden, at Maderanertal and St. Gotthard, Uri, and in the Binnental, Valais in Switzerland.

Although the number of localities yielding really large anatase crystals is limited, the mineral provides a rich source of variety for the micromount collector. Photos of crystals no more than a few millimeters tall, from almost a dozen localities, beautifully illustrating a range of color, luster, crystal shape, zoning, and some interesting associated minerals were presented by Henderson (1992) in an aptly titled article "Anatases I Have Known."

Figure 74. A tiny but sharp crystal of anatase from the type locale, Bourg d'Oisans, France. The crystal is about 2 mm tall. *RJL1492*

Figure 75. A sharp, transparent amber anatase crystal with an equant habit, about 4 mm wide on gneiss, illustrating a classic Alpine-cleft specimen, from Kollergraben, Switzerland. *RJL1956*

Brilliant metallic blue-black crystals, associated with colorless quartz or white adularia, have been found in quantity at several places in Norway since the late 1960s, and although some deposits have been largely exhausted or closed to collecting, they must be regarded as classic anatase occurrences. Noteworthy locales include Slidre in the Valdres area; Sjoa; Kragero; Gudbrandsdalen; and Hardangervidda (Cook 2002). Collectors should be aware that specimens from Hardangervidda (vidda = high plateau) might show a variety of locality names such as "Hordaland, Norway." The correct locality has been given as Matskorhae, Ullensvang Statsallmenning, Hordaland Fylke, Norway (Griffin et al. 1977).

Figure 76. Lustrous black anatase crystals about 1 cm long on white adularia, from Hordaland Fylke, Norway. *RJL2450*

Figure 77. Blue-black striated anatase crystals on a 4-cm quartz crystal from Statsallmenning, Hordaland Fylke, Norway. *RJL 851*

Figure 78. Large doubly terminated anatase crystal about 15 mm tall on matrix from Hordaland Fylke, Norway. *RJL1493*

The quartz crystal deposits in the Subpolar Urals, Russia, although perhaps more noteworthy for large tabular brookite crystals, have also yielded nice examples of anatase on either smoky quartz crystals or on schist. The anatase crystals are typically a few millimeters long, but some up to 12 millimeters have been reported (Burlakov 1999).

Figure 79. Anatase crystals to about 6 mm long, on schist from Dodo, Russia. The crystals are partly transparent yellow, with areas of blue-black. *RJL2349*

Figure 80. Sharp transparent to steel-blue anatase crystals to about 4 mm, scattered on a transparent smoky quartz crystal from Dodo, Russia. *RJL1570*

Fine, dark, equant crystals have become plentiful in recent years from several localities in Pakistan. According to Dudley Blauwet (*pers. comm.* 2008), "It seems that the anatase started to appear at least a year after the brookite started. I have not gone to Baluchistan; the location is quite a long and now very dangerous journey from Peshawar, so I have no idea of the local geology, except from the specimens. We have had a hard time getting exact locale information and it started as Dalbundi (sp.) and ended up as Kharan. Both are larger cities in the area and someone told me that the location is roughly somewhere in between the two and about 5-6 hours by jeep from Kharan." It is interesting to note that the time lag between when the brookite became available and when the anatase specimens entered the trade in quantity suggests that most came from a different area or deposit. However, specimens are occasionally found with excellent brookite and anatase crystals growing side by side.

Figure 81. Anatase crystals to about 6 mm, on matrix with flattened quartz crystals from Kharan, Baluchistan, Pakistan. The crystals illustrate both elongated and truncated bi-pyramidal habits. *RJL2975*

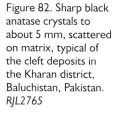

Figure 82. Sharp black anatase crystals to about 5 mm, scattered on matrix, typical of the cleft deposits in the Kharan district, Baluchistan, Pakistan. *RJL2765*

Anatase occurs along with rutile and brookite at Magnet Cove, Arkansas. It is found at several locales in North Carolina, including: in mica schist at Yates Brooks farm, Lattimore, Cleveland County; and on quartz at a cleft-type deposit at the Shingletrap Mountain mine, Montgomery County.

Figure 83. Blue-black anatase crystals to about 3 mm on matrix from the Mid-State quarry, Magnet Cove, Arkansas. *RJL2674*

Figure 84. Tiny black anatase crystals on mica schist from Cleveland County, North Carolina. *RJL3294*

Figure 85. Dark blue striated anatase crystal about 2 mm tall in mica, from the Yates Brooks farm, Lattimore, Cleveland County, North Carolina. *RJL3428*

Figure 86. Iron-stained, rough crystalline quartz with minute black metallic anatase crystals, from the Shingletrap Mountain mine, Montgomery County, North Carolina. *RJL3430*

Pale brown anatase crystals are found in a range of interesting morphologies in Bahia, Brazil; many are sharp but fairly simple tabular shapes, whereas others exhibit elongated "sawtooth" structures indicating fluctuating growth rates. As noted above, crystals are sometimes found in various stages of alteration to rutile.

Figure 87. Several thumbnail-sized several anatase crystals from Brazil, illustrating an interesting range of habits.

Figure 88. Brown bi-pyramidal anatase crystals about 8 mm long, perched on a heavily included quartz crystal, from Cuaiba, Gouveia, Minas Gerais, Brazil. *RJL3257*

Figure 89. A complex anatase crystal from Gouveia, Minas Gerais, Brazil, about 70% replaced by golden-brown needles of rutile. Specimen is about 2 cm tall. *RJL3331*

Pseudomorphs and Alteration Products

Pseudomorphs of anatase after perovskite as a result of weathering have been described from the Salitre II carbonatite, Minas Gerais, Brazil (Banfield and Veblen 1992) and at the Catalão-I carbonatite, Goias, Brazil (Pereira et al. 2005). Good examples are also found at Perovskite Hill, Magnet Cove, Arkansas; this material was originally described under the name *hydrotitanite* by Koenig in 1876 (Palache, Berman, and Frondel 1944).

Anatase replacing titanite crystals up to about 1 centimeter can be found in Corral Canyon, Dixie Valley, Churchill County, Nevada, associated with albite. Pseudomorphs after titanite are also found at Roanoke County, Virginia, and at the Jones zircon mine, Henderson County, North Carolina (Vance and Doern 1989). The site in Virginia comprises an allanite-bearing pegmatite near the crest of the Blue Ridge, where crystals in sizes up to 4 centimeters are found. The crystals are wedge-shaped, very typical of titanite, and are light yellow and opaque. Freshly broken surfaces have a porcelainous luster (Mitchell 1964). At Green River, Henderson County, North Carolina, yellow earthy material having a soft, friable texture was found as an alteration product after titanite crystals and described in the mid-1800s under several names (*xanthitane, xanthotitane, xanthotitanite*). This material was shown to be identical with anatase (Palache, Berman, and Frondel 1944).

Brookite

Brookite is typically found in cleft-type deposits in gneiss and schist, associated with anatase, titanite, orthoclase, rutile, quartz, hematite, calcite, chlorite, and muscovite. It is sometimes formed by contact metamorphic processes and in some hydrothermal veins. Brookite is also a common detrital mineral.

Figure 90. Tan anatase forming a sharp pseudomorphic replacement of a 9-mm perovskite, from Perovskite Hill, Magnet Cove, Arkansas. *RJL3072*

The original description of brookite by Levy (1825) was based primarily on crystals from Wales, supplied to him by James Sowerby, and in part on crystals from Dauphin, France, supplied by a Mr. Turner; Levy recognized that both samples were the same species. According to modern practice, it would perhaps be appropriate to recognize the two places as co-type locales. Brookite from the first location was originally referred to as being from "near Snowdon" by Sowerby (Levy 1825), but later Sowerby confirmed the location as Prenteg. It was also incorrectly called Fronolen by Greg and Letsom (1858) and elsewhere described as "8 miles from Snowdon, between Beddgelert and Tremadoc."

Interestingly, the exact locale is known today with great precision and was described in detail by Starkey and Robinson (1992), who suggest the correct label would state: Twll maen grisial (cave of the crystals), Fron Olau, Prenteg, Gwynedd (Caernarvonshire), Wales. The site has yielded fine crystals to three centimeters across, associated with quartz, albite, anatase, and other minerals, often partially enclosed in late-stage carbonates. The Alpine cleft-type mineralization is attributed to a very low-grade regional metamorphic event that altered the host dolerite (Starkey and Robinson 1992).

Figure 91. A small tabular red-brown brookite crystal about 3 mm tall on quartz, from the classic locality at Prenteg, Wales. *RJL2960*

Large tabular brookite crystals are found in quartz-lined clefts at the Dodo deposit, Subpolar Urals, Russia. The crystals are extremely thin and brittle. On average they are about 0.5 millimeter thick and 1 x 2 centimeters across; rarely crystals up to 12 centimeters long are found but these are usually already fractured, either by tectonic movements, repetitive freezing and thawing of ice, or from blasting at the mine (Burlakov 1999, Cook 2003b).

Figure 92. A large thin brookite crystal, about 3 centimeters tall, with a 1-cm long colorless quartz crystal attached, from Dodo, Subpolar Urals, Russia. *RJL1985*

Magnet Cove, Arkansas, may be considered the premier American brookite locale. Thick tabular crystals larger than one centimeter and rutile paramorphs after brookite up to six centimeters have been found. The brookite at Magnet Cove typically contains some Fe and Nb, as well as up to one percent vanadium, and is therefore black (Howard 1999). Thin tabular brookite crystals are found at several locales in California, including Hale Creek, Trinity County (Dunning and Cooper 2000) and near Georgetown, El Dorado County (Hadley 2000).

Figure 93. Small, thin tabular brookite crystal about 15 mm tall, from Luzenac mine, St. Pierre de Broughton, Quebec, Canada. *RJL3332*

Figure 94. Dark euhedral brookite crystals to about 1 cm on corroded milky quartz from Magnet Cove, Arkansas. Specimen is about 10 cm tall. *RJL3119*

Figure 95. A small brookite crystal, about 5 X 5 mm, perched on a corroded milky quartz crystal, from Magnet Cove, Arkansas. Front view shows the generally square outline and heavy surface figures on the crystal. *RJL399*

Figure 96. Side view of the specimen shown in the previous photo, showing thick tabular habit and that the forms on the side of the crystal are somewhat smoother than the termination. *RJL399*

Figure 97. Another specimen from Magnet Cove, Arkansas, about 3 cm wide, showing an intergrown cluster of gray, terminated quartz crystals, thickly covered by equant, 5-mm brookite crystals. *RJL2673*

Figure 98. A less-common association from Magnet Cove, Arkansas: a steel-gray brookite crystal about 7 mm wide on striated cubic pyrite crystals. *RJL3369*

Figure 99. A large specimen with a long history: this cabinet-sized piece of massive to crystalline quartz, with a lustrous, black 15-mm brookite crystal was accompanied by several old calligraphic labels when it was sold during the dispersal of the collection of the Philadelphia Academy of Sciences. *RJL3396*

Figure 100. Another unusual association piece from Magnet Cove, Arkansas, consisting of hundreds of 3-4 mm euhedral brookite crystals cemented by a massive iron oxide-rich matrix. *RJL3101*

Spectacular crystals from Pakistan first emerged on the market in late 2004. As discussed above for anatase, the indicated locale for most of this material is "Kharan, Baluchistan, Pakistan" and in fact there appear to be a number of cleft deposits in the general area that are producing specimens (Dudley Blauwet, *pers. comm.* 2008). Many early specimens were rather simple, thin tabular brownish crystals, sometimes containing dark, hourglass-shaped zones. Zoning of this type was noted many years earlier in brookites from Ellenville, New York, and from Tavetsch, Switzerland, and attributed to localized concentrations of niobium (Frondel, Newhouse, and Jarrell 1942). As more

Figure 101. A thin tabular brookite crystal about 1 cm tall on colorless quartz, from Kharan, Pakistan. The dark "hourglass" zoning can be seen clearly. *RJL2739*

specimens appeared, a greater variety of habits have been observed, ranging from metallic black striated crystals, thickened in the middle with thin sharp edges, to thick, deep red tabular crystals having terminations similar to those of the type material from Wales.

Figure 102. Superb reddish-brown tabular brookite crystals on matrix with colorless quartz, from Dalbandi, Pakistan. Crystal in foreground is about 12 mm tall. *RJL2807*

Figure 103. Very sharp black, striated, tabular brookite crystals with a submetallic luster, from Kharan, Pakistan. *RJL2974*

Figure 104. Brown tabular brookite, nicely formed with a termination similar to that seen in specimens from Wales. This unusually thick crystal is associated with colorless quartz, from Kharan, Pakistan. *RJL 3365*

Figure 105. Tabular brookite from Kharan, Pakistan, fairly typical of the locale, associated with bright golden rutile needles, some encased in colorless quartz. *RJL3000*

Figure 106. A thumbnail-sized specimen with brookite crystals to about 12 mm tall on colorless quartz, from Kharan, Pakistan. *RJL2782*

Recently, deep brown tabular crystals from 2 to 4.5 cm long were found in an Alpine cleft-type occurrence at Ikalamavony, Fianarantsoa Province, Madagascar, along with a few 8-mm anatase crystals (Moore 2008). A discussion of some additional brookite locales in the United States and worldwide can be found in Cook (2003b).

Figure 107. Sharp transparent brown brookite crystals to about 4 mm long, scattered on sandstone from Monte Bregaceto, Gorssetto, Liguria, Italy. *RJL2131*

References

Banfield, J. F., and D. R. Veblen 1992. Conversion of perovskite to anatase and TiO₂ (B): A TEM study and the use of fundamental building blocks for understanding relationships among the TiO_2 minerals. *American Mineralogist* 77:545-57.

Bernard, J. H., and J. Hyršl 2004. *Minerals and their localities.* Prague: Granit.

Beurlen, H., C. de Castro, R. Thomas, S. B. Barreto, and L. E. Prado-Borges 2004. Strüverite and scandium bearing titanian ixiolite from the Canoas pegmatite (Acari – Rio Grande do Norte) in the Borborema pegmatitic province, NE-Brazil. *Estudos Geologicos* 14:20-30.

Brammall, A., and H. F. Harwood 1923. The occurrence of rutile, brookite, and anatase on Dartmoor. *Mineralogical Magazine* 20 (100): 20-26.

Brown, D. L., and W. E. Wilson 2001. Famous mineral localities: The Rist and Ellis tracts, Hiddenite, North Carolina. *Mineralogical Record* 32 (2): 129-40.

Burke, E. A. J. 2006. A mass discreditation of GQN minerals. *Canadian Mineralogist* 44: 1557-60.

Burlakov, E. V. 1999. The Dodo deposit, Subpolar Urals, Russia. *Mineralogical Record* 30 (6): 427-42.

Buseck, P. R. and K. Keil 1966. Meteoritic rutile. *American Mineralogist* 51: 1506-15.

Cairncross, B., and S. Moir 1996. Famous mineral localities: The Onganja mining district, Namibia. *Mineralogical Record* 27 (2): 85-97.

Černý, P., B. J. Paul, F. C. Hawthorne, and R. Chapman 1981. A niobian rutile – disordered columbite intergrowth from the Huron Claim pegmatite, southeastern Manitoba. *Canadian Mineralogist* 19: 541-48.

Cook, R. B. 1985. Famous mineral localities: the mineralogy of Graves Mountain, Lincoln County, Georgia. *Mineralogical Record* 16 (6): 443-58.

Cook, R. B. 2002. Connoisseur's Choice: Anatase, Hardangervidda, Ullensvang, Norway. *Rocks & Minerals* 78: 400-3.

Cook, R. B. 2003a. Connoisseur's Choice: Rutile, Graves Mountain, Lincoln County, Georgia. *Rocks & Minerals* 78: 112-6.

Cook, R. B. 2003b. Connoisseur's Choice: Brookite, Dodo mine, Subpolar Urals, Russia. *Rocks & Minerals* 78: 192-4.

Crook, T., and S. J. Johnstone 1912. On strüverite from the Federated Malay States. *Mineralogical Magazine* 16 (75): 224-31.

Dachille, F., P. Y. Simons, and R. Roy. 1968. Pressure-temperature studies of anatase, brookite, rutile and TiO2-II. *American Mineralogist* 53: 1929-39.

Deer, W. A, R. A. Howie, and J. Zussman 1962. *Rock-Forming Minerals*, Vol. 5 Non-Silicates. London: Longman's.

Delametherie, J.-C. 1797. *Theorie de la Terre*, Second Edition, Vol. 2: 268-71.

Dunning, G. E. and J. F. Cooper 2000. Brookite from Hale Creek, Trinity County, California. *Mineralogical Record* 31 (4): 341-43.

El Goresy, A., M. Chen, L. Dubrovinsky, P. Gillet, and G. Graup 2001a. An ultradense polymorph of rutile with seven-coordinated titanium from the Ries crater. *Science* 293: 1467-70.

El Goresy, A., M. Chen, P. Gillet, L. Dubrovinsky, G. Graup, and R. Ahuja 2001b. A natural shock-induced polymorph of rutile with α-PbO₂ structure in the suevite from the Ries crater in Germany. *Earth and Planetary Science Letters* 192: 485-95.

Evseev, A. A. 1993. The South Urals: A brief mineralogical guide. *World of Stones* 1: 31-5.

Evseev, A. A. 1996. Worldwide mineralogy: A sketch of an exposition. *World of Stones* 9: 33.

Flinter, B. H. 1959. Re-examination of "struverite" from Salak North, Malaya. *American Mineralogist* 44: 620-32.

Frondel, C., W. H. Newhouse, and R. F. Jarrell 1942. Spatial distribution of minor elements in single-crystals. *American Mineralogist* 27: 726-45.

Gaines, R. V., H. C. W. Skinner, E. E. Foord, B. Mason, and A. Rozenzweig 1997. *Dana's New Mineralogy*, Eighth Edition. New York: John Wiley & Sons.

Goldschmidt, V. 1913. *Atlas der Krystallformen* Vol. 1 [see Facsimile Reprint in Nine Volumes (1986) by the Rochester Mineralogical Symposium].

Goldschmidt, V. 1922. *Atlas der Krystallformen* Vol. VII [see Facsimile Reprint in Nine Volumes (1986) by the Rochester Mineralogical Symposium].

Greg, R. P., and W. G. Letsom 1858. *Manual of the mineralogy of Great Britain and Ireland.* London:John van Voorst.

Griffin, W. L., T. Garmo, H. Løvenskiold, and A. Palmstrøm 1977. Anatase from Norway. *Mineralogical Record* 8 (4): 266-71.

Hadley, T. A. 2000. Brookite and anatase from near Georgetown, El Dorado County, California. *Mineralogical Record* 31 (4): 345-48.

Henderson, W. A. 1992. Microminerals: Anatases I have known. *Mineralogical Record* 23 (1): 88-90.

Hochleitner, R., N. Tarcea, G. Simon, W. Kiefer, and J. Popp 2004. Micro-raman spectroscopy: a valuable tool for the investigation of extraterrestrial material. *J. of Raman Spectroscopy* 35 (6): 515-18.

Horn, M., C. F. Schwerdtfeger, and E. P. Meagher 1972. Refinement of the structure of anatase at several temperatures. *Zeitschrift für Kristallographie* 136: 273-81.

Howard, J. M. 1999. Brookite, rutile paramorphs after brookite, and rutile twins from Magnet Cove, Arkansas. *Rocks & Minerals* 74: 93-102.

International Mineralogical Association 1962. International Mineralogical Association: Commission on New Minerals and Mineral Names. *Mineralogical Magazine* 33: 260-63

Jamieson, J. C., and B. Olinger 1969. Pressure-temperature studies of anatase, brookite, rutile, and $TiO_2(II)$: A discussion. *American Mineralogist* 54: 1477-81.

Koivula, J. I. 1987. The rutilated topaz misnomer. *Gems & Gemology* 23 (2): 100-03.

Kouznetsov, N. 2001. Recent mining of minerals and gems in Russia by Stone Flower Company. *Mineralogical Record* 32 (1): 42.

Levy, M. 1825. An account of a new mineral. *Annals of Philosophy*, Vol. IX: 140-42.

Mandarino, J. A., and V. Anderson 1989. *Monteregion treasures: The minerals of Mont Saint-Hilaire, Quebec.* Cambridge: Cambridge University Press.

Middlemost, E. A. K. 1985. *Magmas and magmatic rocks: An introduction to igneous petrology.* London: Longman.

Mitchell, R. S. 1964. Pseudomorphs of anatase after sphene from Roanoke County, Virginia. *American Mineralogist* 49: 1136-39.

Moore, T. P. 2008. What's new in minerals. *Mineralogical Record* 39 (3): 233-45.

Nagelschmidt, G., H. F. Donnelly, and A. J. Morcom 1949. On the occurrence of anatase in sedimentary kaolin, *Mineralogical Magazine* 28: 492-95.

Palache, C., H. Berman, and C. Frondel 1944. *Dana's system of mineralogy*, Seventh Edition, Vol. 1. New York: John Wiley & Sons.

Pauling, L., and J. H. Sturdivant 1928. The crystal structure of brookite. *Zeitschrift für Kristallographie* 68: 239-56.

Periera, V. P., R. V. Concieção, M. L. L. Formoso, and A. C. Pires 2005. Alteration of perovskite to anatase in silica-undersaturated rocks of the Catalão-I carbonatite complex, Brazil: A raman study. *Revista Brasiliera de Geociencias* 35(2): 239-44.

Platt, R. G., and R. H. Mitchell 1996. Transition metal rutiles and titanates from the Deadhorse Creek diatreme complex, northwestern Ontario, Canada. *Mineralogical Magazine* 60: 403-13.

Posch, T., F. Kerschbaum, D. Fabian, H. Mutschke, J. Dorschner, A. Tamanai, and T. Henning 2003. Infrared properties of solid titanium oxides: Exploring potential primary dust condensates. *The Astrophysical Journal Supplement Series* 149: 437-45.

Prior, G. T., and F. Zambonini 1908. On strüverite and its relation to ilmenorutile. *Mineralogical Magazine* 15 (8): 78-89.

Ranade, M. R., A Navrotsky, H. Z. Zhang, J. F. Banfield, S. H. Elder, A. Zaban, P. H. Borse, S. K. Kulkarni, G. S. Doran, and H. J. Whitfield 2002. Energetics of nanocrystalline TiO_2. *Proc. National Academy of Sciences* 99, suppl. 2: 6476-81.

Roberts, W. L., and G. Rapp, Jr. 1965. *Mineralogy of the Black Hills.* Bulletin No. 18 of the South Dakota School of Mines and Technology, Rapid City, South Dakota.

Scrivenor, J. B. 1912. Notes on the occurrence of strüverite in Perak. *Mineralogical Magazine* 16 (76): 302-3.

Spry, P. G., and N. L. Scherbarth 2006. The gold-vanadium-tellurium association at the Tuvatu gold-silver prospect, Fiji: Conditions of ore deposition. *Mineralogy and Petrology* 87: 171-86.

Starkey, R. E., and G. W. Robinson 1992. Famous mineral localities: Prenteg, Tremadog, Gwynnedd, Wales. *Mineralogical Record* 23 (5): 391-99.

Steele, I. M. 1975. Mineralogy of lunar norite 78235; second lunar occurrence of P21ca pyroxene from Apollo 17 soils. *American Mineralogist* 60: 1086-91.

Tollo, R. P., and S. E. Haggerty 1987. Nb-Cr-rutile in the Orapa kimberlite, Botswana. *Canadian Mineralogist* 25: 251-64.

Vance, E. R., and D. C. Doern 1989. The properties of anatase pseudomorphs after titanite. *Canadian Mineralogist* 27: 495-8.

Wise, W. S. 1977. Mineralogy of the Champion mine, White Mountains, California. *Mineralogical Record* 8 (6): 478-86.

Withers, A. C., E. J. Essene, and Y. Zhang 2003. Rutile/TiO_2 II phase equilibria. *Contributions to Mineralogy and Petrology* 145: 199-204.